THE
LAST SHEPHERDS

TEXT BY DAVID OUTERBRIDGE
PHOTOGRAPHS BY JULIE THAYER

A STUDIO BOOK · THE VIKING PRESS · NEW YORK

Text Copyright © David Outerbridge 1979
Photographs Copyright © Julie Thayer 1979
All rights reserved
First published in 1979 by The Viking Press
625 Madison Avenue, New York, N.Y. 10022
Published simultaneously in Canada by
Penguin Books Canada Limited

Library of Congress Cataloging in Publication Data
Outerbridge, David.
   The last shepherds.
   (A Studio book)
   Bibliography: p.
   1. Shepherds—Europe.   2. Sheep—Europe.
3. Shepherds—Pictorial works.   4. Sheep—Pictorial works.
I.   Thayer, Julie.   II.   Title.
SF375.5.E85095   636.3′08′300922   78–20858
ISBN 0–670–41891–9

Page 141 constitutes an extension of this copyright page

Printed in the United States of America
Set in Baskerville

CONTENTS

Since a moment in prehistory when man began to domesticate animals there have been herders. Herders have played an important role in almost all cultures and until today this role has continued. There are herders of reindeer on the arctic subcontinent, of alpaca and llama in the Andes, of sheep, cows, goats, yak, camels, buffalo, and even of turkeys. Together with hunters, fishermen, and farmers, they have been the providers of food.

Herders share common knowledges. They must know the medicinal remedies for any sickness among their animals, and the habits of predators and how to protect their herds from them. Most of all, they must know the weather and terrain so that the animals can find food and water and, on occasion, shelter. The search for pasture is central in their lives.

The evolution from hunter to herder was a simple one: during a hunt a baby lamb—or kid or fawn—was discovered. Instead of being slaughtered it was brought back and given to a woman to suckle. By the time it was weaned, several weeks later, it had become tame in human company. Afterwards, it continued to stay nearby until one day, driven by an unnamed instinct, it returned to the ground on which it had been born. There, during the rutting season, it was bred. The following spring it returned to the place of the people with its offspring.

Over time, man learned that he could control the freedom of the animals by the construction of brush barriers or by tethering. These controls could be relocated when a patch of grass was eaten off.

The practice of releasing a flock of ewes to be bred by wild rams in the hills continues to this day. And in some places women still suckle orphaned wild animals. In fact, there has been surprisingly little change in the ways of herding through time. Until now.

The herder's role is suddenly coming to an end. Agribusiness has found that large-scale ranching is more effectively handled by bringing pasture to flock in the form of cut and processed grasses. The shearing of animals has become mechanized, and cheese-making is increasingly centralized in large plants. Airplanes and poisons are used in the control of predators; and, finally, what residual herding duties remain are done by hired hands.

As the traditional role of the herder passes into history, so too will the techniques, superstitions, and routines that governed a way of life. This book is a glimpse into that life as seen through the days of several shepherds.

When we think of a herder the image that most likely comes to mind is that of a sheep herder, and there is reason to believe that the shepherd was the first herder. According to early history sheep were the favored animals of sacrifice. There are remains of domesticated sheep in the dwellings of much of the early civilizations, from Mesopotamia to the Irish crannogs. But whether the shepherd came first or not, he enjoys a unique position in Western culture: guardian of the soul, symbol of a pure life. Of all the traditional occupations of man the shepherd is most visible; visible not in physical presence, for his habitat is a faraway hill, but in art, religion, literature, and legend.

The Bible is our earliest reference:

And Adam knew Eve his wife; and she conceived, and bore Cain, and
     said, I have gotten a man from the Lord.
And she again bore his brother Abel. And Abel was a keeper of sheep,
     but Cain was a tiller of the ground.
And in process of time it came to pass, that Cain brought of the fruit of
     the ground an offering to the Lord.
And Abel, he also brought of the firstlings of his flock and of the fat
     thereof. And the Lord had respect unto Abel and his offering.
But unto Cain and to his offering he had not respect.

The Bible can be traversed on a chronology of shepherds —Jacob, David, et al.—until, in the New Testament, the Lord himself becomes shepherd and we his flock.

Western art is resplendent in its portrayal of shepherds. They appear as part of idyllic landscapes, and in renditions of ancient myth, and within Christian iconography.

Early in man's ability to abstract, the shepherd was isolated as symbol of a first need: nourishment, for he was a provider of food. Thus came Pan, ancient god of the flocks. He is known to Homer who made the error of ascribing to him the current meaning of the word "pan"— "all." Thus Pan came to be described as god of all living things. But he was not that, for in the earlier language of Greece his name means "he who feeds." From that meaning the symbol grew over the centuries. In Pan, simple herders personified health among their animals. To an even earlier hearth Pan only meant that there was food to eat.

Pan, shepherd of Arcadia, was friend of the nymphs. As herders do, he moved among the hills with his flocks, rested at noon, had no care for agriculture. From classical times he has been known primarily for two accomplishments: his ability to instill fear in man and beast and cause stampedes, from which comes the English word "panic"; and his love of music, and creation of shepherd's pipes—also called panpipes or the syrinx. The legend of this musical instrument of shepherds is this:

Pan was excited by the beauty of the maiden Syrinx and wanted to have her. She, cherishing her maidenhood, fled into a marsh to hide, and begged the nymphs to save her. Responding, they turned her into a patch of reeds. Poor Pan, desolate, clutched a few stalks to himself and blew sadly into them. To this day the panpipe is the instrument of shepherds from such diverse places as the Pyrenees, South America, and Rumania.

During the classical period of Greece and Rome the epic immortalized shepherds as heroes and gods. Miltiades Xsilouris, the shepherd of Crete in this book, grazes his sheep on the slopes under Mount Ida, where Zeus was said to have been reared. On another Mount Ida, near Troy, the shepherd Anchises also kept his sheep, and was one day visited by Aphrodite. Mortals who flirted with gods did not often have happy endings, and Anchises would have kept to his sheep. Aphrodite, however, had lust in her loins; and who was to refuse the goddess of love? From this dalliance Aeneas was born to become defender of Troy. Alas for the shepherd Anchises, his concern had not been misguided. He was crippled by a lightning bolt thrust at him by Zeus.

On this same hillside Paris, perhaps the most famous of all shepherds, also lived. Shortly after his birth a prophecy told of his eventual destruction of Troy. To prevent that Hecuba and Priam, his parents, left him on the hillside, exposed, to die. However, Paris was discovered and raised by a shepherd, and grew to be a young man of great beauty. One day, while watching his sheep, he was told to judge a contest between three goddesses: Hera, Athena, and Aphrodite. The prize was a golden apple inscribed "to the fairest." Each promised her specialty to Paris. Aphrodite made the winning offer of the most beautiful woman in the world for his wife. Historically, the consequences were immense. From his pastoral life on a Trojan hill Paris went to Sparta to claim his prize, Helen. Unfortunately, she was already the wife of Menelaus. With her abduction the Trojan War began.

What the epic has not detailed for us in the daily life of Anchises or Paris we can learn from Homer's description of another ancient herder, the Cyclops, Polyphemus. The routine of his work is not very different from that of Xsilouris today:

We climbed, then, briskly to the cave. But Kyklops
had gone afield, to pasture his fat sheep,
so we look round at everything inside:
a drying rack that sagged with cheeses, pen
crowded with lambs and kids, each in its class:
firstlings apart from middlings, and the "dewdrops,"
or newborn lambkins, penned apart from both.
And vessels full of whey were brimming there—
bowls of earthenware and pails for milking. . . .

When he came
he had a load of dry boughs on his shoulder
to stoke his fire at suppertime. He dumped it
with a great crash into that hollow cave,
and we all scattered fast to the far wall.
Then over the broad cavern floor he ushered
the ewes he meant to milk. He left his rams
and he-goats in the yard outside, and swung
high overhead a slab of solid rock
to close the cave. . . . Next he took his seat
and milked his bleating ewes. A practiced job
he made of it, giving each ewe her suckling;
thickened his milk, then, into curds and whey,
sieved out the curds to drip in withy baskets,
and poured the whey to stand in bowls
cooling until he drank it for his supper.

There is one shepherd in the writings of ancient Greece who has great importance in the development of Western literature. His name is Daphnis.

Throughout Greece, then as now, the summer sun burned spring vegetation. The withering of grasses was mourned annually. Theocritus, a Greek poet of the third century B.C., symbolized this death in Daphnis. His eulogy elevates Daphnis into a personification of spring's passing:

Now you may bear violets, you bramble bushes; bear them, you thistles, and let the lovely narcissus spread its foliage over the juniper. Let all things be changed, and let the pine tree bear pears, since Daphnis dies, and let the stag drag down the dogs and let the screech owl from the mountains contend with nightingales.

Thus Daphnis becomes nature, and, from spring, the symbol of youth. A shepherd's life is taken as the representation of perfection, a pure and uncorrupted existence. By extension, the shepherd also becomes a metaphor for another poet, or even, in death, a close friend. Virgil uses the death of Daphnis as an allegory of Caesar's death.

Theocritus called his poems "idylls," which gives us our meaning of the word. The literature that uses the shepherd in this way is called "pastoral" from the Latin word meaning shepherd, or less frequently "bucolic" from the Greek word of the same sense. There have been pastoral elegies such as Milton's "Lycidas"; pastoral fiction, the flower of which is Longus' sensual *Daphnis and Chloe;* pastoral drama from Shakespeare and others; and volume upon volume of pastoral poetry. The latter reached its zenith during the Elizabethan era if we use profusion as a measure. By this time the shepherd is used in a myriad of ways. Sometimes he is located in Arcadia, sometimes in the native surroundings of the poet. Daphnis, Pan, and other deities dance in and out of these lines, which ultimately become ridiculous. An example:

He borrowed, on the working days, his holly russets oft;
And of the bacon's fat, to make his startups black and soft;
And lest his tarbox should offend he left it at the fold;
Sweet grout, or whig, his bottle had as much as it would hold;
A sheave of bread as brown as nut, and cheese as white as snow;
And wildings, or the season's fruit, he did in sup bestow;
And whilst his piebald cur did sleep, and the sheephook did
    lay him by,
On hollow quill of oaten straw he piped melody.

(William Warner, from "Albion's England")

Ben Jonson wrote of Pope's pastoral poetry: "It seems natural for a young poet to initiate himself by pastoral, which, not professing to imitate life, requires no experience."

Whether the verse was good or bad the message is, nonetheless, recurrent: that the life of a shepherd was somehow "better."

Gives not the hawthorn bush a sweeter shade
To shepherds looking on their silly sheep
Than doth a rich embroider'd canopy
To kings that fear their subjects' treachery?
O yes, it doth! a thousandfold it doth!
And to conclude, the shepherd's homely curds
His cold thin drink out of his leather bottle,
His wonted sleep under a fresh tree's shade,
All which secure and sweetly he enjoys,
Is far beyond a prince's delicates,
His viands sparkling in a golden cup,
His body couched in a curious bed,
When care, mistrust, and treason waits on him.

(*Henry VI*, Part 3, Act II, Scene 5)

This romanticization of the herder's life is part of the delusion under which young people today think it is a life-style to adopt. "Flocking to the flocks," reported *Time* on the burgeoning enrollment in shepherd schools throughout western Europe. In France, especially, shepherding is a boom occupation in terms of the number of applicants. Graduation does not seem to correlate with longevity in the profession. In most places the independent herder exists less and less; the newcomer to the activity does not persevere. It is a life too far different from the dream. As one lifelong shepherd in the Cévennes remarked: "The young don't want any part of it. It is tranquil, but very hard. There are no vacations. My work stops December 31 and starts again January 1."

The sweet visions must be tempered by the reality. It has always been a hard life, pastoral poesy notwithstanding. That is not to say it is a life without pleasure. The pleasures *are* there within the life and are sufficient if one does not look beyond. They lie within a self-contained system and are not externally produced.

"I did always like sheep," said one shepherd. "Some did say to me that they couldn't abide shepherding because of the Sunday work. But I always said, somebody must do it; they must have food in winter and water in summer, and must be looked for, and it can't be worse for me to do it."

What is a shepherd? Some livelihoods require a closeness to nature, but the shepherd *is* nature. He is as much a part of it as the animals he tends and the predators he fights. The dependence is complete. As much as any animal and more than any man, he is absorbed into the cycles of life and death, the flow of time. There are no days off, but rest within each day. He knows the movements of weather and responds to them by altering his behavior to fit. Unlike the fisherman who hangs up his nets and goes home, the farmer who puts away his scythe, or the hunter who carries home his quarry, the shepherd is called by each setting sun, each rising sun.

"We want to go out and have fun, but the animals never have holidays —you have to take them out every day. I would rather work in a factory. You get paid every week and you have insurance." So commented the granddaughter of a Sarakatsani herder to a *New York Times* columnist last year.

"Isn't life better now, isn't it easier?" she turned and asked of her aged relative. "Not for me," he replied, and walked out of the house and returned to his sheep.

Some shepherds raise sheep for meat, others for cheese, others for wool. Some are transhumants, some are nomads, some are sedentary. Some bring their sheep in at night, others let them graze. Some must check their flock daily, others cannot.

The first domesticated sheep were quite unlike the purebred animals we see today. They were small and wild, covered with rough hair. Wool, as a sheep by-product, developed through selective breeding late in history. The indigenous sheep of Scotland, for example, was not the blackface but a short scrubby animal which survives today only in the Orkney Isles, where it is valued for its ability to survive, well nourished, grazing on seaweed along the shore.

14

Today there are one billion sheep in the world being tended by someone. They range from fat merinos that provide good meat and soft wool to the karakul whose black curly pelt provides the material for "persian lamb" raiments. There are sheep that will give birth to twins or triplets, sheep that can survive brutal winters, hothouse sheep, sheep that need lush pasture, sheep that do well on arid landscape.

The shepherds of this book span the varieties within their work. The modern sheep rancher has not been included. Some of his work, in particular lambing, is still tended to in the traditional way; but most else is new. Shearing is done by a special group of men who work suspended in harnesses that move on overhead tracks. They clip with shears powered from a central rotary unit. Feed-lot systems are used for fattening, trucks and rail for transport. The number of sheep in a herd is huge. The herder, whether a musterer on the Australian Outback or a Basque in Idaho, is a hired hand on contract.

The brink of technological sheep farming is evident in the Scottish Borders. Plentiful rainfall, adequate pasture, and a moderate winter climate fostered extensive early herding here. By the twelfth century monasteries had already organized hill grazing for large flocks. The sheep were raised for wool which was then exported to Flanders to be woven. The shepherds, or herds as they are called in the Borders, today raise a few select breeds that are suited to the terrain and develop a heavy carcass quickly. Cheviots, blue-headed Leicesters, Suffolk rams for crossing. On the "out-bye," or hills, the blackface does best in the harsher environment.

This is fenced land, and the sheep are moved from field to field for maximum use of the land. Hay is harvested and fed to them in winter. Turnips are grown to be used for food before lambing. Supplementary food in the form of processed pellets is also used. The shepherd lives in a nearby farmhouse and returns there from the fields for his meals. Tractors, and other mechanized farm equipment, are available. It could be called sheep farming, the man a farmer, except for this: the work has not changed. It is still a 365-day job, and during lambing the day runs to twenty hours.

At the other extreme are the shepherds who must always move in search of pasture. The word "nomad" comes from the Greek *nomas,* meaning roaming for pasture. The Bedouin, Masai, Afghans, Tartars, and Yakuts still move with their animals across the land. As they may seldom, or never, return to the same ground, they move in families. They know nothing of arable land and live in a world of sand, rock, ice, and snow. They must endure savage weather and will march hundreds of miles to find adequate feed for their animals. They live in tents made of skins which they tan with sour milk and smoke, or yurts made of bark stitched with horse hair. Their only food is what their animals provide. They buy almost nothing: a few essentials are obtained through barter.

In the cold of Siberia frozen mares' milk is cut in chunks and used as money. The nomadic herders are seldom part of the national culture and, although they can represent a sizeable percentage of the population as in Afghanistan, they are outsiders. T. E. Lawrence in his *Seven Pillars of Wisdom* described them thus:

Shepherds were a class apart. For the ordinary Arab the hearth was a university, about which their world passed and where they heard the best talk, the news of their tribe, its poems, histories, love tales, lawsuits and bargainings. By such constant sharing in the hearth councils they grew up masters of expression, dialecticians, orators, able to sit with dignity in any gathering and never at a loss for moving words. The shepherds missed the whole of this. From infancy they followed their calling, which took them in all seasons and weathers, day and night, into the hills, and condemned them to loneliness and brute company. In the wilderness, among the dry bones of nature, they grew up natural, knowing nothing of man and his affairs; hardly sane in ordinary talk; but very wise in plants, wild animals, and the habits of their own goats and sheep, whose milk was their chief sustenance.

If he moves seasonally, the herder is called a transhumant. It is a new word, invented to describe the practice, but the definition must encompass many variations. The move itself is called the transhumance. For some herders of Pakistan it will be a trip from an altitude of 2,000 feet to one of 14,000, a journey made in stages with some gardening en route. For the Lapp and other arctic herders it means the extremes of wintering in conifer forests where the temperature remains —50 degrees for long periods, and summering on the tundra where the air is gray with mosquitoes.

For most of the shepherds of the Mediterranean basin—Greece, Italy, France, and Spain—the transhumance is a semiannual movement between two locations. Even in the faraway Hebrides it was so a millennium earlier. Everywhere, it is a moment of celebration and festivity. Among the Basque, for example,

From first light this morning, the air has been filled with the sound of a thousand bells.

It is a time of great occasion, because on this day in May, the sheep begin their long ascent from the valley floor to the peaks of the Pyrenees, there to graze upon the rich grass until snowfall.

All the country lanes have been crowded with tiny flocks, each starting from its farm at hours appointed by tradition. It is tradition, too, that every flock must pass through the village on its way to the high mountains.

The sheep are shining white from the spring rains and from the brushings they have received in the hectic week of preparation before. Their curling wool hangs down like a blanket almost to the ground. They have patrician noses and their gracefully curved horns have been decked with ribbons by the children. Each flock is distinguished from another by splashes of powdered paint, red and yellow and blue and orange, at different places on the white backs.

When they pass through the village, the leaders of the flock come first with staidly nodding heads and huge gourd-shaped bells that make a deep and hollow sound befitting their age. The lambs who frolic on the edges of the flock have collars shaped from green willow, and bells that make a tinkling sound.

At the heels of the sheep are the little shepherd dogs with quick movements and eyes that peep brightly from behind long fringes of hair.

And lastly come the shepherds, leading mules piled high with bedding and white sacks with provisions they cannot find on the mountain, wine and coffee and tobacco and sugar and salt.

The shepherds are clean-shaven for the occasion. Some are dressed in berets and suits, as if they were going to church. Others wear the costumes of times before, jerkins made from the pelts of lambs, woolen cloaks against the cold of the mountains, and legs bound up with leather thongs. Long black umbrellas with cane handles are hooked behind on the collars of suits and cloaks, and all of them carry a wooden staff worn smooth by the touch of their hands.

The shepherds walk as if they were in a parade, as indeed they are, because their flocks must bear the appraisal of the villagers who line the streets. The shepherds walk straight and proud, with the remote faces of men who have learned to live in solitude. They are not without their vanity, however, and the villagers' response to the quality and beauty of each flock is reflected in the shepherds' demeanor.

The procession has lasted until darkness. Now that it is over, the mood of the village has changed. In every country lane, in every house that touches upon a meadow, in every dawn and sunset, the sound of bells is gone.

If the shepherd has a wife and family, they do not move with him but remain at the principal residence even though he may be gone for six months. The shepherd may take a few rabbits or fowl with him, and will plant a small garden of greens at the next location. In some parts of Europe "home" is in the valleys and the shepherd goes up into the mountains only after the grasses begin to dry up. For others, the reverse is true: they leave their homes at the approach of winter and find pasture in the more temperate lowland. This was true of Europe's most important system of transhumance, the Castilian.

The transhumance was an extremely important part of the Spanish economy as early as the thirteenth century, and the routes taken from the mountains were themselves pre-Roman. The organization of sheep-owners, the *Mesta,* was so powerful in Castile that shepherds were exempt from military conscription, could carry arms (to ward off wolves and gypsies), and did not need to pay tolls on the roads as they passed through different jurisdictions. Three million sheep moved annually on a long transhumance of 350 miles. The journey took a month. En route through the mountains in the spring, the sheep would be sheared in long sheds where teams of more than one hundred shearers did the work. The wool was then sold, usually still "in the grease," and packed for export.

Any transhumance presupposes complicated internal and external structures and weighty institutions. In the case of Castilian wool, it involved towns and markets like Segovia; Genoese businessmen who bought up wool in advance and, like the Florentines, possessed vats where the fleeces could be washed, not to mention the Castilian agents for these big merchants, the transporters of the bales of wool, the fleets that sailed from Bilbao for Flanders (controlled by the Consulate of Burgos), or the consignments sent off to Alicante or Malaga, destined for Italy; or even to take an everyday detail, the indispensable salt which had to be bought and transported to the grazing lands for the flocks.

Today, the transhumance is much the same, but the commerce that gave it international significance has gone.

"There ain't nothing dumber than a sheep except the man who herds 'em." [Saying from the United States West.]

The shepherds of the United States are located for the most part in the West. There are Indian herders in the Southwest. In the northwestern states are the vast flocks of the lamb industry. For many the cycle is still transhumant.

A great number of the shepherds are Basque and they arrive by jet from the Pyrenees, a contract in their pocket. They have not herded before and undergo a fast apprenticeship. After three years they return to their homeland and will not be shepherds again. Their newly earned dollars will enable them to become entrepreneurs.

The use of Basques as herders goes back to the days of the Gold Rush when many of them arrived to seek their fortune. They came via South America where they had been stockmen. To this day they continue to provide the manpower for flock tending. It is not, however, altogether a story of pride, for the Basque herders were that for just one reason: to make money and get out. Driven by opportunism they gave little concern to overgrazing of the land. This misuse of broad sections of the country was such that Congress passed laws specifically to regulate grazing patterns.

Long before these laws, however, another pressure was on the western sheepman: the cattleman. The latter was justifiably critical of the way sheep cropped the range, and attempted to keep them out. The conflict was perennial. Gun battles were fought over whether sheep or cattle were going to be on the land. In fact, the worst family feud in the history of the United States was not fought in the Kentucky hills, but in Arizona over this issue.

It was called the Tonto Basin War, or Pleasant Valley War. The valley had been used by Graham family cattle for generations. No sheepman had ever brought his "woolies" into the valley. None had dared to try. Until one day a ranchhand on the Graham range saw, incredibly, the "maggot-like hordes" appearing. Galloping back to the Graham homestead, he announced: "The Tewksburys are driving sheep over the rim of the Mogollons."

20

The feud ran for years and before it was over every member of the Graham family was dead, the Tewksburys with severe losses as well. The residual bitterness is such that almost one hundred years later Zane Grey, interested in researching the subject for a book, spent three years there before any local person would speak of the happening.

They say of a pig that everything is used except his squeal; for the sheep, too, only the bleat is lost.

Wool is used in many ways. The coarsest grade is used as mattress stuffing, the next for carpet weaving, then for cloth yarns. The best wool, used for knitted goods (once from St. Kilda, today still available from the Shetlands), is not sheared but pulled with a penknife from the fleece, perfect tuft by perfect tuft.

Sheepskins were the original garb of shepherds and also formed his tents and sacs. This is still true in some places, but sheepskins have more generally become fancy merchandise.

The slats, or woolless pelts, are tanned and made into fine leather products. Formerly slats also were made into parchment which was used as translucent window material, for writing, and for academic diplomas.

The horn is used for making buttons, the intestines for the finest-quality surgical gut; from the lining of the stomach comes rennet. In addition, lanolin and tallow are taken from the sheep.

Until recently, the children born of herders became herders themselves. It was the only way of life they knew and the apprenticeship started very early. It probably happened thus:

His earliest memory of herding went back to when he spent the day alone with the family's five sheep. It was midafternoon—the shadows already long on the hills—and time to bring the animals back to the stone and bramble enclosure. He was six and had done this on other days, that he could still vaguely remember, with his mother. But on that day he was alone. It was very different: for the first time the sheep wanted to go somewhere else. He ran to show them the right way, but they bunched up and moved farther away. He tried to catch the last one—it kept looking at him—but somehow it didn't work. He had left the bottom of the valley, and the stream through it seemed a long way off. On the other side was home.

He sat down on a rock, angry. There was no sign of help. He *must* get them back. Then he remembered the nuts. In his pocket was a small handful picked from a tree and carried for this purpose. He pulled them out, as his mother did, and rattled them. The sheep turned to the familiar noise. He had become a herder.

The initiation into full responsibility varied from place to place. Among some herding societies initiation required the stealing of a neighbor's sheep. In others the moment came when, within the family structure, one inherited a portion of the flock. These were the prerequisites for ownership, independence. However, the routine protection of the herd against the elements and predators had its own rites of passage.

It was at this point that he had his vision of death. It came with the first clear consciousness of exhaustion. He just suddenly saw how men and sheep perished. He realized the power of the elements. He heard the tearing whine of the wind, looked at the whirling flakes, and in a moment's curious detachment recognized his absolute impotence. He had got heated rescuing the sheep, and now the wind cut into him. It also slightly choked him with its searching soft fist. He had to gulp now and then. There was a sense of strain on the ribs of his chest. The snow was flattening against the back of his head, behind his ears. To hear the wind's whine was to hear the scream in it and beyond it, the writhing scream of a sweeping invisible blade.

Worn out, a man lay down—and never woke. It was known. The sheep searched for shelter, found it, and then the snow covered them. But the sheep had to be dug out. Lost here and covered deep, they too would perish. No searchers would find them or him till they found the dead bodies in the wash of the thaw.

In most countries the sheep are marked in some manner for identification purposes. One method is to notch the ear. In some cultures a shepherd has many different ear notches to distinguish age and breeding within his flock. In one region a mark is burned into the side of the face. Today, most shepherds use a blotch of color on the rump. This used to be done with tar and was applied to give a cross or other specific design. Tar has been replaced by dye which in France is carefully applied in a motif or initials or both.

The markings, however, are really only a convenience. Even unmarked, a shepherd can always recognize his sheep. Within a flock of several hundred, for example, he can tell you this sheep was caught in a fence two days last year, that one over there had twins.

In some parts of Iran, it is accepted that if a herder cannot recognize his sheep within someone else's flock, they are not his.

Of all the predators the wolf dominates. No matter what country, the stories of wolf against sheep and shepherd abound:

"An ill shepherd doth often feed the wolf."

"Spare my kids, O wolfe, and spare my mother-goats, nor harm me for that I am so little to tend so large a flock."

"It is hard to keep the wolf full and the wether whole."

In France, the sheep dogs were armed with huge collars of iron spikes to fend off a wolf attack. In Hungary, the komondor became a favored guard dog for the flock because its densely matted coat was almost impervious to a wolf bite. Special lanterns that glowed like sheep's eyes were used by shepherds to hunt the wolves, and elaborate signal systems were employed to warn each other of a wolf attack.

The wolf is gone from Britain and the Mediterranean basin. The fox, however, is still abundant and takes a share of lambs each year. In Scotland, special terriers are used to locate and enter their cairns and kill the young pups. In the United States coyotes and lynx pursue the sheep; eagles carry off lambs in many countries; crows and gulls will attack a sheep that is hurt, peck out its eyes, and kill it. The shepherd must still guard against all of this.

In France the vipers which frequent much of the pasturage are another menace. If a sheep is bitten on the tongue it will die. Otherwise, the shepherd cuts the bite with his knife and urinates in the wound. This will usually save the animal.

Shepherds have traditionally used simple remedies: in France angelica for wounds, juniper for digestive disorders, ironwort and mallow for other ailments. Before the development of hardier breeds in Scotland a mixture of tar and butter would be spread on sheep to protect them from the winter cold. If a sheep develops an abscess in its mouth or eye, Yves Hébrard, a French shepherd, cuts a slit in the ear and inserts a small roll of leather. This draws the infection out of a vital organ into a superfluous section of skin that can be removed.

For all their hardiness in poor country, sheep are vulnerable to a fatal malady if they eat too much legume. It is called bloat, or blast, and creates a disorder in the first stomach which will kill the animal unless the gas is released. In *Far from the Madding Crowd,* Thomas Hardy describes the shepherd, Gabriel Oak, at this task:

Gabriel was already among the turgid, prostrate forms. He had flung off his coat, rolled up his shirt-sleeves, and taken from his pocket the instrument of salvation. It was a small tube or trochar, with a lance passing down the inside; and Gabriel began to use it with a dexterity that would have graced a hospital-surgeon. Passing his hand over the sheep's left flank, and selecting the proper point, he punctured the skin and rumen with the lance as it stood in the tube; then he suddenly withdrew the lance, retaining the tube in its place. A current of air rushed up the tube, forcible enough to have extinguished a candle held at the orifice.

The costume and equipment of the shepherd vary according to topography and climate. The cape is quite universal, and serves not only as protection by day but also as bedding at night. The capes of the Cévennes are made of canvas and lined with a heavy wool plaid. The capes of Hungary are cut more like the sarapes worn by the Indian herders of the Andes. The Sardinians also have wool capes, blankets really, with just one clasp. In Scotland protection used to include a plaid belted cape worn over bare thighs and wool leggings. In the Landes, as in many parts of Asia, the cloak was made of sheepskin. Underneath were worn a vest and leggings of the same material, though thighs and feet were bare. In this costume, according to one report, the shepherds differed little in appearance from the animals whose hides they wore.

Aside from the cape, the shepherd's costume often looks surprisingly formal for the work. The Sardinians wear suits of hardy chestnut-colored velvet; the Highlanders tweed jackets and trousers, and a coat of durable gabardine if it rains. Some sort of staff is used everywhere. Partly it is used as a walking stick in covering the terrain, partly to control the sheep. In much of France, it is a straight staff to which is attached a length of leather belting making a simple whip that is effective in dividing or corralling groups of sheep. In the Borders of Scotland, the staff is about five feet long with a crook. It is made of hazel and the crook is either carved out of wood or ram's horn and then countersunk. In Greece, the staff has a crook made by one bend of the wood while green.

Almost all the shepherds carry a sac of some sort. Usually, they are made of canvas, sheepskin, or leather.

The special equipment in different parts of the world is varied. Huge cloth umbrellas are used by shepherds in Provence and are specially made. To alleviate the fear of lightning the tines and shaft are made of wood, and so that it may serve as a simple tent in the field, the shaft is made long to form a good angle from the ground. In the Landes, a marshy wet land, the shepherds used to wear high stilts. Not only did this keep them dry, but it also allowed them to watch over their charges more easily. The stilts were very tall—the footrest six feet above the ground—and on them a walking man could pace a trotting horse. In the Scottish Highlands, where the sheep are spread out over mile upon mile, shepherds carry a long glass or binoculars. In Sardinia, where sheep stealing is endemic, the herder carries a rifle.

Finally, dogs have been used by herders for centuries and special breeds have been developed as sheep dogs. Undoubtedly, the most renowned of these breeds is the border collie which is used throughout Scotland, Wales, and England.

"Herding is nothing to do and all day to do it in." (Saying from the United States West.)

Exactly what does a shepherd *do* all day long? During lambing, the busiest time of year, work is continuous and there is scarcely time for sleep. For the shepherd who makes cheese, his day is a sixteen-hour one for more than half the year. For all herders, annual chores include shearing, dipping, castrating, and marking. Some lambs will need to be nursed on a bottle; some sheep will need surgical or medicinal treatment. The produce of the flock—whether it be wool, meat, or cheese—must be marketed. Walls, tents, fences must be repaired. Food must be grown.

Different terrains command the day in different ways. The crofter, Duncan McDonald, can check only a part of his herd in a full day's walk on the moors. The Cevenole herder returns his sheep to the fold each evening. He does this partly to protect them and partly to be able to gather the manure which is a valuable commodity.

When the sheep graze and there truly is time to rest, some shepherds knit, some carve new collars for the sheep or wooden mugs and spoons with appropriate motif. The Basque herders build stone monuments that dot pasturage both in the Pyrenees and the western United States.

And relax. As one observer noted a hundred years ago, shepherds were the only ones in a village who had time to dream. In that solitude, which is not loneliness, they see nature. Comparing town and rural life, Vita Sackville-West wrote:

Book-learning they have known.
They meet together, talk, and grow most wise,
But they have lost, in losing solitude,
Something,—an inward grace, the seeing eyes,
The power of being alone;
The power of being alone with earth and skies,
Of going about a task with quietude,
Aware at once of earth's surrounding mood
And of an insect crawling on a stone.

One shepherd on time: "There is always time for a drink of wine." And, "There never is enough time in the day except when one is forced inside by the weather. Then one is bored."

Two sides of the same coin.

Nomadic herders are likely to be family men. For others, there is little time at home, and if the shepherd is a transhumant he is away half the year. Many crofters never marry. Romance is not very visible, the "sweetest sport of shepherds" and other pastoral wishes notwithstanding. The time is given elsewhere. Yet, when shepherd did meet shepherdess "and if the two herds became mixed, the people in some corner of the field did likewise. And if another herder, looking from afar, saw two flocks together without being tended, he knew what was happening."

To the tourist who chances upon a shepherd's hut there would be wonder: why are there sacs of salt and of sheep dung in the "living room"? Why is the bed only rugs on the floor? What are the ointments and powders on the disorganized shelves? This tourist, walking into a shepherd's stone hut in Crete, or a windowless croft in the Hebrides, or a herder's wagon on the Sussex downs, would have thought: they must be very poor to have such little houses.

But for the shepherd who lives his life out-of-doors it is space enough, and friendly space at that.

In the far, far distance the sound of bells is like a faint chorus without melody but nonetheless musical. A single tone that comes now softer, now louder—like a living thing—modulated by the wind and the contours of the terrain.

Closer, but still beyond calling distance, the noise bubbles and could be confused with the sound of a brook working its way through rocks and pools. It is a pleasant, hollow sound.

Then, still closer, amid the general noise one can hear individual bells— some particular in their high clink, others in their depth, and still others distinguished by their clear melodious sweetness.

Only when one is with the flock does the general noise reveal its composition: a medley of hundreds and hundreds of individual tones, no two alike, perhaps five hundred pitches all within one octave.

And when the flock has passed from within hearing distance one hears the bells still. Is it imagined? Perhaps. Yet one cannot not *hear them. Even in the midst of conversation (or in city noise) the sound continues on for some time.*

Perhaps the first image of a shepherd that comes to mind is that of a boy on a lonely vigil in an Alpine meadow, or of someone sitting on a rock, playing his flute, or, perhaps, of a Basque herder, wearing his beret.

There are still sheep in the Alps and the Pyrenees, but the way of herding is changing. Throughout most of the Mediterranean basin shepherds are forced to move semiannually in order to find adequate pasture for their flocks. This was always a journey by foot. Now, however, the sheep are transported by triple-tiered trucks.

Only in the Cévennes are traditions still somewhat intact. But even there they are changing. Ethnologists come to the Cévennes to record the changes in the area while various government agencies try to maintain things as they were. One misguided department even started to reintroduce wolves onto the grazing lands, recalling the shepherd's traditional battle with the predator; fortunately, the program was soon discontinued.

Of course things *will* change in the Cévennes, the old ways will pass. At one time the sheep of a shepherd who had died would not be decorated for the transhumance for a period of five years. Although the period of mourning is still observed, nowadays no decorations appear in the sixth year either. It takes half a day to put them on, to no purpose. Yves Hébrard has always decorated his sheep: "Pom-poms are a tradition, and because I am old I do it. *Voilà.* Otherwise it is like a regiment without music: it wouldn't work."

The flocks throughout this area were once blessed by the local priest before they left for the mountains. The benediction took place in the village square and was followed by a pagan ceremony, during which great piles of herbs were gathered and burned and the sheep guided through the ashes. It was believed that the ritual would protect the sheep against disease and, like many ancient customs, there was probably some medical truth to it. Even today, though the ashes are no longer used, Hébrard hangs a large bunch of the herbs—spurge— under the arch of the sheepfold at lambing time.

From the Camargue to the Pyrenees, through the land of the Languedoc, there is a common climate and similar terrain. It is a country of olive trees, vineyards, and fruit orchards. The poorer land is given to grazing of sheep. It is ground covered with a scrubby growth, but it is also aromatic. In addition to the box and thistles which predominate, there is a profusion of thyme, lavender, and oregano. One cannot help but crush the herbs underfoot. From these meadows come the shepherd's medicines and his barometer, the *chardouille*. This flower with large petals opens at the approach of dry weather, closes with the advent of rain. Walls and doors throughout the area are decorated with the flower.

The annual migration has been going on for a long time. It was approximately four thousand years ago that the weather changed in the Cévennes and the grasses started burning in summer. The wild animals of the region were the first to start a seasonal move to a more temperate altitude. As man settled in the area and began keeping animals, he too found it necessary to move for pasture. He found his route to the mountains by following the paths of the wild animals. These paths are still used today, worn by the migrations of centuries.

It is almost time to head for the mountains. Around St.-Hippolyte the grapevines are thriving, but the grass is becoming dry. Some shepherds have already begun their trip. There is no specific date for departure: "One looks at the weather, looks at the grass, one gathers one's things and departs."

In the spring, meetings are held nonetheless among the shepherds to discuss dates and routes to take in order to avoid the confusion that can occur if two flocks bound for different mountains simultaneously arrive at a crossing of paths. Fees are negotiated for those herds that will be placed in another shepherd's care for the summer, and rendezvous for the combining groups are established. These meetings take place in the cafés of the towns and the discussions are carried on in a festive manner over wine.

The weather looks promising for the next few days and Hébrard makes preparations for departure. Now that spring is over the lambs are weaned and strong. The sheep have been sheared. (This is no longer done by hand shears; Hébrard hires professional shearers for the job.) A few sheep, still with full fleece, have been saved for a special honor: they will wear red, yellow, and blue pom-poms for the transhumance. They are the best-looking of the flock and include an undue percentage of black sheep. The latter are selected by Hébrard because he has a special feeling for them; he believes they are bullied when young. These sheep he clips by hand, leaving three mounds of wool along their backs to which the decorations can be attached.

Special, large bells are worn by a few selected sheep. The deep
tone of the bells of the transhumance carries well across the route and
serves as a guide for the shepherd who must know the whereabouts
of the lead sheep. In some parts of France these bells are fastened
to the sheep by a yoke.

In addition to these bells worn by a few, there are regular bells
worn by many. Each region of France has its own variations in size and
design, which evolved quite naturally as local foundries executed their
commission. Today, there are only three foundries left and they service
all the communities but still maintain the regional shapes at their forges.
Because these bells are handmade there are differences in tone, even
among those made for the same area. When a shepherd comes to buy
them he will listen to each in order to determine which will sound
harmonious with the other bells on his flock.

There is a difference of opinion among shepherds as to whether
each sheep should have a bell. Some feel it is wrong as it agitates the
flock. Others place bells on almost the entire flock so they may hear if a
sheep has strayed. Still others would like to have that many bells but
cannot afford to buy them. Hébrard places bells on almost all of his
sheep.

Throughout the Languedoc the collars that are used with the sheep
bells are made of wood. In former times the collars were emblems of the
herder's craftsmanship. Each made his own and carved the completed
collar with intricate geometric designs, usually adding his initials and
sometimes borders of clover, fleur de lis, or other pattern. Many
shepherds still make their own collars, although they can also be bought.
Usually they are painted rather than carved.

After he has completed decorating and belling the special sheep,
Hébrard smears a handful of mud on their backs. It is mud brought
down from the mountain the previous autumn for this purpose. He
has always done it, it brings good luck.

The departure. It is still dark when Hébrard awakens. Two sons have arrived to help. Neither is a shepherd but for the next three days, and again in the fall, each will do what he can to make the trip easier. Roland, a neighbor who occasionally tends sheep and enjoys the walk to the mountains, arrives. He will walk in front of the flock during the transhumance. An old shepherd, Amadé Puel, also arrives. The previous autumn he came on the morning of the descent and asked to lead the sheep the first twenty kilometers of the trip.

"But you are not well," Hébrard reasoned.

"Better to die in front of a flock than in bed."

"Yes, it is true! Go, then."

Hébrard and the others go to the pen of rams and harness a leather apron under each. In this way the rams can travel with the flock, yet they are unable to mate with the ewes. The apron serves as a simple and effective contraception, until the descent in October when the rams will serve their function and impregnate the entire flock during the three-day journey.

The goats, which have a special importance for the shepherd of the Cévennes, also make the trip. They provide milk which will be made into cheese, and they will also be used to suckle orphaned lambs. The goats are not necessarily penned with the sheep; sometimes in the mountains they share the shepherd's abode while the dogs sleep outside. Being intelligent animals, the goats are aware of their state of privilege and frequently press for treats.

There is a rushed meal of bread and cheese, a swallow of wine. Hébrard ties his bootlaces. He whistles for the dogs, gates are opened, and without any apparent organization twelve hundred sheep are suddenly trooping down the road. Dawn is coming to the land.

The distance between St.-Hippolyte and le Vilaret, where Hébrard will spend the summer, is one hundred kilometers. The trip will take three days. For the first few miles the procession travels over a paved road, and a few cars are encountered. Those coming toward the flock park and wait for it to pass. A bread truck comes from behind. With loud cries Hébrard starts moving ahead through the sheep using his whip like a propeller to clear a lane. The truck follows close behind. The sheep re-form as soon as it passes. Eventually it passes through, and the driver waves a greeting as he pulls ahead.

By eight-thirty the sheep have left the road and have started the climb. At first the trail leads through woods and at times only a small section of the file is visible. But the air is full of the sound of bells, and as they climb higher the valley echoes with the clamor.

Throughout the Cévennes other shepherds have looked at the weather and are departing. One of them, René Carrière, *maître-berger,* master shepherd, will start with his own flock. All during the day other flocks will join until, by nightfall, he will have fifty-seven different flocks—totaling almost five thousand animals—entrusted to his care for the summer.

In the little village of St. Martial four small roads lead into a dusty square shaded by stately old trees. One side of this square is formed by a café which has a balcony set with tables. Just after 10:00 A.M., preceded by the sound of bells, a flock of seven hundred sheep arrives. They move into the shade under the trees and rest. The shepherd moves to the balcony to enjoy a rest, as well. He orders a Pernod and some lunch. A few minutes later a second troop arrives. Then a third and a fourth. Over two thousand sheep are packed into the tiny place, and four shepherds and village friends are packed onto the balcony overlooking them. Big loaves of bread are cut, cheese and wine are passed. The air smells of Gauloise and sheep. In a few minutes the whole group will leave for a predetermined rendezvous with Carrière in the evening.

After a walk of six hours, with one pause for grazing, Hébrard stops for lunch. His mother and sons have been awaiting him above a small stream in the hamlet of Colognac. It is a lengthy rest for shepherd and sheep. Hébrard and Roland sit with the family and eat the food prepared by Hébrard's ninety-one-year-old mother the day before. There is pâté of wild boar, chicken, sausage, cheese, olives, bread, and fruit. There is wine.

By nightfall of the first day Hébrard has reached an abandoned shelter. It is in the clouds and the air is wet. A bonfire is made for some warmth and a little cooking, then it is time to sleep.

Before dawn the shepherds eat a cold breakfast and go out to gather the sheep off the slope. The trek continues, Roland leading, Hébrard bringing up the rear. He has two dogs, one of which is a pup in training. Working with the older dog and with stern words from the shepherd, the pup will learn a great deal during these three days.

It is a steady, steep climb throughout the morning. At lunchtime the family is there again, having driven by car to an intersection of the trail. It is a shorter meal this time as the weather that had looked so promising changes.

The afternoon track is narrow and the sheep are spread along a distance of almost a mile. Color slowly leaves the mountainside, only the yellow-flowering gorse remaining particular. It grows darker and darker, then green. What was a rumble of thunder becomes individual cracks and booms. Lightning dances into the slope. Hébrard opens his umbrella and walks on.

It rains, then it is a parade of drenching showers within the rain. It grows colder. Soon in the downdrafts it is hailing—white balls the size of marbles pelt the convoy. It is still more than a two-hour walk before the trail widens and there is a place where the sheep can rest and feed. They are moving slowly now, and Hébrard sits for a moment on an outcropping of rock by the path. He looks old. Almost immediately he nods into a thirty-second sleep, then awakens and starts walking again. It is wet and cold.

At an intersection with another path he meets a former shepherd who has come out to greet him. They walk together for a kilometer, talking. People join him like this periodically along the route, and he enjoys the company.

He walks on in the rain. The rear of the flock is the place of most responsibility during the transhumance. Sheep that are straggling or climbing away in search of grass must be brought back into line. Hébrard calls to his dogs, talks to the sheep, whirls his whip and keeps them moving. Someone offers him a bottle of wine. He pauses, uncorks it, and drinks under his umbrella. Then he recorks the bottle, returns it, and continues.

It is four in the afternoon and still raining steadily when they arrive at the resting place, Bonperrier. Twenty years ago eighty thousand sheep passed through this spot. Today the number is ten thousand. Roland and Hébrard stand under their umbrellas and allow the animals to graze for two hours. The sheep steam in the rain.

Hébrard is worried: the next shelter and pasture are one and a half hours farther away, and part of the path is very narrow with a long drop alongside. In the dark and with lightning he is afraid some sheep could be lost. They move on, sodden.

When he was young, he remembers, there were no shelters and one wrapped himself in a great cape and huddled for the night under an umbrella. He still carries the cape for protection from cold during the descent in the fall.

The cape and the umbrella, plus the shepherd's sac, are the basic equipment. In 1976, Charles Coste, who for seventy-one years made the equipment of shepherds, died at his bench at the age of eighty-five. His widow still keeps a box of letters—most of them written in a semiliterate patois—in which shepherds ordered their needs. In all of France his death left only one other such artisan, and he died in June, 1977.

It is the third and last day of Hébrard's transhumance. Dawn breaks and the troop moves on through the last of the mountain paths. Hébrard stops for lunch at Cabrillac, near the point where the mountains give way to the high plateaus of the Causse Mejean. Again, it is a long rest but without family this time.

They move on.
The sun disappears behind powerful clouds of evil burnish. It is broad treeless country and the sheep make their way in a many-fingered configuration of files.

It has been raining hard for two hours when the destination is in sight: le Vilaret, an abandoned eighteenth-century farmhouse with a cluster of stone outbuildings, is rented by Hébrard during the summer grazing period. Rent for pasture is still largely paid in sheep droppings which the shepherd rakes together and bags each morning after the flock has left the fold.

For the last mile Hébrard changes place with Roland to lead his animals home. The pace has quickened as men, dogs, and sheep drive on to shelter. A son has arrived ahead to light a fire. The sheep are pushed rapidly into the barns and penned. Hébrard moves indoors, folds his umbrella, hangs up his sac, sheds an outer layer of soaked clothing, sits down by the fire. He is chilled and does not feel well. He drinks a small glass of wine. It was terrible weather for the transhumance this year, he says.

He is tired, but home.

Jean Pibarot, who lives about thirty miles west, in the hamlet of Agonès, was for many years an assistant shepherd. Now, he has his own flock which he tries to enlarge a little each year. That is hard to do, because a small flock of ewes will produce few lambs. To meet expenses he is forced to sell so many lambs that few—perhaps none—can be kept for herd maintenance. He must pay rent for winter pasture, buy food for the dog, and a new ram if possible; there is medicine to be purchased, shots and dips required by the government must be administered. Last year Pibarot was forced to sell his lambs before the spring transhumance. It is always a bad time to sell because the coming summer is a period of good food, when the lambs fatten rapidly. It is a season that takes little money, only time.

At a certain point the break-even will be reached, perhaps at 250. Then Pibarot will be assured of a sufficient crop of lambs both for market and to retain for breeding stock.

This year he has taken charge of other flocks for the summer pasturing and their passage to it. This provides a fee. Instead of looking after two hundred sheep, he is caring for five thousand. But he is a good shepherd and they will be well attended.

"Since I was three I have always followed the sound of the bells. Even then, if I was missing my mother always would know where to find me. My only regret in life is that my wife comes from the city and so cannot share with me the enjoyment of the way I spend my days.

"I never like to stop for lunch until the sheep are tranquil and I will have an hour without disturbance. Today, the sheep are restless and I am going to move them to another field. Today, I will not be eating lunch before two o'clock. But it is all right. A shepherd always has time."

In the autumn, after the return from the mountains, Pibarot can be found every morning with his sheep in the woods. It is the season of acorns, which they love. And, he says, the dealers believe the acorns give the meat good flavor, so they will pay more for them. The sheep cannot stay long on the acorns, however, or the acidity of the tannin will disrupt their digestion. But for a time each day one can find Jean Pibarot by listening for the sound of sheep bells in the woods.

Within a radius of Roquefort the business of sheep is given to milk rather than meat. To make even a meager living here, one must produce fifty liters of milk a day. This is then shipped to St. Jean du Bruel to be converted into a fresh cheese, and then further transported to the Roquefort caves where it matures.

Only four families live in the tiny community of Laupiettes, cut into a steeply terraced hillside. The younger members of these families have long since left for work in the cities. A few hens scratch alongside the road, and two huge pigs live in a pen bedded with clean dried ferns. A cow was once kept as well, but now the pasture cannot support both sheep and cow. Each family still tends a small flock of ewes that produces under twenty liters of milk a day.

Laupiettes is a village of old people and old grass.

Scotland's history is not easily traceable in the town of Mallaig on the coast of the western Highlands. The town itself goes back but a hundred years, for it was not until the terrible Clearances that people were forced to flee to this harsh landscape. Today it is a wealthy town made rich by catches of herring, bottomfish, prawns, lobsters, and cockles. Twice a week huge flats of fish and crans of "silver darlings," as herring are called, are auctioned off to buyers who will deliver them across Britain.

Sheep are everywhere . . . on the quay, clipping the sod of the cemetery grounds, in the churchyard, in gardens, and backyards, outside pubs, and in the middle of the town's one street. And, being sheep, they find amusement in disrupting traffic and upsetting garbage cans. The people of Mallaig erect ever higher fences—of wood, wire mesh, or steel —but to no avail. The sheep are just as likely to be contained within as they are to be locked out.

At one town meeting when the subject of sheep came up, as it invariably does, one indignant resident complained that the morning after he had gone to the expense and trouble of installing a new four-foot cyclone fence he looked out his window to admire his handiwork, but was startled to see a sheep at the moment of aerial transit. The narrator paused, listening for sympathy and eyeing the audience for culpability. Silence. Presently, Donald McDonald, his mouth twisting into a smile, spoke in a voice of exaggerated pride: "Well, if she could scale four feet she'll be ours."

Donald and Duncan McDonald are brothers and are heirs to the Crofting Act that decrees the perpetual right of sheep to graze on this land. Because the sheep were there before the town existed, under law the sheep cannot be rezoned.

*Mallaigvaig, little Mallaig.
Small stone house backed into what in Gaelic is called "hill of the
goats," facing the wild and ever-changing Sound of Sleat.*

*Duncan McDonald, the end of so many histories. To walk for a day on
these hills and moors is to see the histories, some in visible remains, some
in the names of places, some which come alive in the mind watching
his work.*

Son of a crofter, born on this coast. Although there are a dozen crofters
in Mallaig, if one applies the legal meaning of the word, there is only
one: Duncan. Even brother Donald does not quite qualify, because he
does not tend sheep and till the land.

A crofter, by definition, does not own his land but has tenure over
it as long as the annual rent is paid. A typical croft is composed of a
few acres of land on which there is a small dwelling with barns or
outbuildings, plus sufficient ground to till for potatoes and other food
crops. A larger piece of land, which can be as much as several thousand
acres, normally goes with the croft. This land is used for grazing.

DUNCAN MC DONALD <<CROFTER>> MALLAIGVAIG

This year Duncan McDonald turned a new piece of land for his potatoes. It is a big patch of many drills and will provide the staple food for the year for himself, his brother and sister, and the dogs. After the potatoes are dug in the fall he stores them in a pit which he digs in the garden, then covers the pit with a thatch made of bracken. Protected in this way from frost the potatoes will last the winter.

Sgur Eireagoraridh, the hill behind the moors where the sheep graze, is named for what happened here one thousand years ago. *Aridh,* Gaelic for shieling, itself a word of Norse origin, means "the place of summer grazing." Hill of the valley of the summer grazing. (Places with names ending in –ary or –shiel which dot a map of this country carry this sense.)

The shieling included a group of small huts of thick stone walls and sod roofs. Here families summered. Boys watched over the pasturing sheep and cattle while women worked with spindle and distaff in preparation for winter weaving. Here crottle and other vegetation were gathered for dyes, clothing was knitted, and men drained land, built walls, and broadcast grasses to harvest.

Land that is now empty except for sheep once held a community. Here, in what are now only piles of rocks, meals were made of whey and cooked grains, and of game on the occasions when it could be caught.

Ancient in origin, the shieling system existed until recently. A traveler in 1786 noted:

About sunset we cast anchor in an open road, at the mouth of the loch, and seeing a decent looking house, with sundry huts at some distance, Macdonald and myself bent our way thither, as if certain of a good reception, of comfortable lodging, and a whole budget of news. When we got to this place, a dead silence pervaded the whole village; the windows and doors of the principal houses were shut; we knocked in vain, nothing that had life was seen or heard from any quarter . . . we were informed by a transient traveller that the people of the village had just gone to the shielings.

The annual summer movement to the shieling was called "the flitting," a word also of Norse origin. The flittings developed from the need to find more pasture.

The shieling system ended in the Borders of Scotland when the raising of sheep was systemized by the monks who already in the twelfth century recognized the lucrative commerce in exporting wool to Flanders. Here, in the Highlands, commerce came much later, but the cheeses of the shielings were always much respected and considered far better food than that which could be made from cow's milk.

It was always poor land. Until this century (and well into it among the outer islands) shieling houses of stone and sod were without windows and chimneys. This was also true of the winter homes. It was the way to conserve heat, but there was an additional purpose. As peat fires burned in a central hearth the smoke rose, held aloft by the heat. Fish and meat could hang there from beams and be preserved. In spring the inner layer of straw thatch on the ceiling, now impregnated with soot, was removed and carefully placed on prepared ground to fertilize the next season's crops.

The sheep of earlier days were a small breed, but they provided the essential cheese and wool for a family. By the end of the eighteenth century, however, the great landowners of the Highlands became more interested in the commerce of wool than in the welfare of their land tenants. It was time to clear the land of people and small stock and introduce a breed of sheep that could generate good revenues. This pogrom was called the Highland Clearances.

Bliadhna nan Caoraich. 1792, Year of the Sheep. The Cheviot, a large and improved breed, was available from the Borders. Together with the black-faced Linton, the introduction of the Cheviot would bring one hundred years of human wickedness and tragedy. "Word of it reached the Highlands in the old way, on the lips of a seer who traveled from township to township, calling a warning: *'Mo thruaighe ort a thir, tha'n caoraich mhor a' teachd!'* Woe to thee, oh land, the Great Sheep is coming!"

The "Clearances" is graphic nomenclature for the program of clearing the land of people, but it can hardly make specific the misery it inflicted on individual families. Burned out of their homes, possessionless, their animals slaughtered, the elderly sometimes buried alive or burned with the house, they were driven north and west to the barren reaches of the coast. There, fish and seaside gardens would provide, Scottish landlords assured themselves. But there was no way so many people could survive without animals or cultivated land.

Into these coastal valleys they came. Many died of starvation when the potatoes failed in 1846; others drowned along the rocky coast in an inept search for fish among the spume-covered rocks. "Mile on mile on mile of desolation / League on league on league without an end," wrote Swinburne.

Some were able to flee the land by sailing for North and South America. Even today, from Patagonia to Oregon, McDonalds, McPhersons, and McClellans are familiar names. Many settled in Nova Scotia (New Scotland) where Highland traditions are still strong, and the homeland is recalled in song:

From the lone shieling of the misty island
Mountains divide us, and the waste of seas
Yet still the blood is strong, the heart is Highland
And we in dreams behold the Hebrides.

Even today the Clearances are not quite forgotten. Bitterness is etched into the culture. In the Church of Scotland the Gaelic readings of the Songs of Praise have been altered to make the Lord a cowherd.

In the 1880s the abuse was finally arrested, and the Crofting Act gave the crofter permanent tenure of his piece of land in exchange for a fixed rent. The original stone fences, houses, and barns from that period are still standing, although the sod roofs have long since broken through the timbers.

Clear edges of long rectangular spaces mark the area where peat was cut. Other lines in the heavy sod attest to the work that was done. Rows of stones set to hold the earth, dug channels to drain land, and the faint contours of cultivation show where potatoes and turnips were once carefully tended. Without ploughs, and without animals to draw them, all earth was dug with spades. Row upon row, drill after drill, turned and formed in the resisting soil by hand. Peat was cut by hand, stacked, and when dried hauled on the back to the croft. Everything turned by spade, carried on oaken trunks of legs.

Today, it is not so different. Sgur Eireagoraridh dominates the landscape as it always did, and the broad land underneath its shadow is the same. The wind on quiet days still blows like an endless current across the braes and down the glens; if it is making the clouds scud overhead it is screaming below. *Cannach,* strange little plants with tufts that look like cotton, are harbingers of spring which comes very late to the Highlands. They used to be known as "the hill's salvation": spring *will* come. For a brief period of time the Hill is green, the bog myrtle and heather bloom, and the gorse is in flower. ("When the gorse is out of flower kissing is out of season," is a saying here. The reason: the gorse is never out of flower.)

Until the modern road and rail systems allowed easy transport of livestock, the vast herds of sheep and cattle being raised in the Highlands presented a considerable problem in terms of marketing. And out of the problem evolved a new breed of entrepreneur, the drover.

The drover was herder, cartographer, trader. It was he who gathered individual flocks, arranged for boat transportation if they were on the islands, then moved them to the fall markets in the south. In a countryside without roads where raiding was frequent droving required consummate skill. Sometimes on the droving path there was a line of wool five to seven miles long; and the one hundred thousand sheep on the move needed water and pasture daily during a trip that lasted for weeks. The places where the drover stopped the procession for grazing were called stances. They can still be located today from their startling greenness caused by the manure of earlier centuries.

To feed both himself and his dogs the drover carried a pouch containing oatmeal which he cooked with water. Or, like the Masai, he would bleed some cattle and then blend the blood with the oatmeal to make a simple black pudding, to this day a traditional dish in the Highlands.

Under the charge of drovers millions upon millions of animals moved across Scotland in this way. Out of the passes of the north, southward along bygone trails, all arrived in time for the famous autumn tryst at Falkirk.

While the hazards and hardships of his calling must fill much of the canvas of any picture of the drover's life, there were some patches of brightness to alleviate the sombre colouring. His wayfaring life taking him far afield, did more than satisfy the love of movement and adventure handed down to him from the generations before. Gossip and talk with other travellers and drovers at inns and wayside meetings cheered his lonely journeys, and with the packsman, the pedlar and the tramp he shared the function of news carrier so dear to country people at a time when news was scarce. His was the excitement of the Tryst, of the bustle and the bargaining; his too, the pride of recounting its every detail to eager listeners back in his Highland glen. If at times these advantages may have seemed to the drover small recompense for his life of hardship, and if he were sometimes tempted to turn his back on the drove road and to prefer his poor croft and scanty holding in Uist, Skye or Lochaber, some other thought may have come to him. Perhaps rough hard men though the drovers were, their memories brought to them as they sat by their peat fires pictures which made them forget their hardships and their sufferings: sunlight and cloud-shadow on the hills of Kintail; gold of the birches in Glen Garry or along the Dee as they took a drove south to the October Trysts; green of lush grass and flags set against the yellow of seaweed as the Islay cattle moved up the shores of Loch Sween; Cruachan with an early powdering of snow; Loch Awe and Loch Lubnaig still on a September morning, or countless other scenes of loch, river, meadow and mountain as they and their cattle passed in the autumn days on their slow journeys through the Highlands of Scotland.

On these broad expanses binoculars are an important tool. "Sometimes you'll see something white on the Hill. You don't know if it is a sheep down or a rock. It is a half-hour walk, possibly to look at a rock, if I don't have binoculars."

Duncan is on the moors checking his sheep. In the far distance, maybe a mile and a half away, he sees some gulls. He looks through his binoculars but nothing seems amiss. "There'll be a dead sheep in the bog, right enough," he comments. Forty-five minutes later he pulls a ewe out of the water. Her eyes and part of her nose have been eaten by the gulls.

The first grass of spring grows near the water, and the sheep, lured by it, fall into the soft bog and cannot get out. Before they are dead the birds are making their meal.

Across most of Scotland the crofters and other subtenants of the land used to pay their rent in kind. For most, this meant the delivery of cartloads of tweeds; from the Hebrides, including famous Harris, it came in boatloads. Bolt after bolt of somber homespuns were dyed with local lichen, the dye set in sheep urine, dried in peat smoke, the wool spun and woven at home. For years the Harris tweed, especially, was sold to gentry and was always identifiable by its pleasant fragrance. Now the wool is washed, spun, dyed, and woven by machine and chemicals, and has lost its characteristic texture and smell.

The origin of all tweed was the tough-fibered wool of the blackface sheep. Across the Highlands, after the sheep are gathered from the hills, the quiet work of shearing is done.

In Mallaig it is a job that takes two months as the soft sound of snipping tells of a single pair of shears doing its work.

The Hill is formidable. Walking steadily, Duncan will reach its
summit in two hours. The best sheep, he says, choose the top where
the grass is fine. To check for sick sheep on the top with its many dips
and crags takes several hours.

During lambing he checks his sheep every day. In a typical year
he will carry ten lambs the four miles back to the croft to be raised on
a bottle. Out of a flock of 450 ewes he will lose fifty to the weather and
the bogs, and as many lambs to foxes. The sheep are blackface; a
Cheviot could not survive on this terrain. If a ewe twins, one will die
if Duncan does not find it in time. There is insufficient grass for the
mother to feed two. Part of the year, when the Hill is covered with
snow, the only food is heather, and "sharp is the tooth that can feed on
the heather" goes a Gaelic saying.

It takes several long days to make one check of the entire flock,
and on days when he is on the Hill Duncan does not bother to bring
lunch. "You start in the morning, but you're all over the top, and by
the time you are home again it will be late."

The sheep are only gathered into the croft twice a year: once to be
sheared and once to be sorted for market. For the gathering Duncan
and his brother row a boat along the coast and then ascend the back of
the Hill, driving the sheep in a homeward direction.

The Hill. Sgur Eireagoraridh. A hard piece of land that makes
its demands on man and beast, yet supports both. Improvements are
not likely to come here; the life is too marginal for anyone to bother
about. Fishing will be the next history of Mallaig. Duncan will be the
last man here to depend on sheep. Yet, for all its austerity, it has been
a good life: "If the Hill would grow just a little shorter I would never
retire."

The Borders have a hard history. England lies just across the River Tweed, and many times the waters have run red as legions of the King in London carried out their punitive expeditions.

Today, on either side of the Cheviot Hills, the soft wet rolling division of territory, shepherds work much the same. One needs a special dictionary to understand herding language here. Hoggs, gimmers, yowes, and cast yowes are names for female sheep at different ages. Rams are tups, and thus a bred ewe has been "tupped." Unlike the Highlands, however, English *is* the language.

Highland shepherds are not given good marks by Border herders: "They're not up 'til dinnertime." To which, of course, the Highlander responds: "Well, they coddle them down there, they do, they do."

It is not important to argue the issue, however. Each is different, still, from herding on the downs of Suffolk or Wiltshire. Each is appropriate for the terrain.

The field is an oblong stretch of rich green grass. Three hundred sheep are scattered across its ten acres. The dog enters the field under a gate and as it does the sheep do not cease eating, but they look up and are aware of its presence. The shepherd walks through the gate, leaving it open, and speaks to the dog. Instantly the dog is gone, flattened by the lowered head and stretched length of a dead run. Down the distance of the near fence line, up the shorter back end of the sloping pasture, then down along the far line moving in along the sheep. The random white dots merge into an edge.

"Come by to me, by-da-me, by-da-me, Nell." The dog reverses its course and races back to the far end, around the edge, and crisscrosses behind the flock. The mass tightens as it moves together except for the sheep along the near side.

"Come away to me, Nell. Guay-da-me, guay-da-me, guay-da-me, Nell." The whizz of black and white is making a parallel line along the lower fence that is very close to the sheep. Now, the entire bunch has a clear white perimeter set against green. One sheep veers away in a nervous revolt. The dog has outrun it in a few seconds, and returns it to the flock with vengeance, nipping a hind leg to accelerate the process.

"Doon, Nell. Doon." The dog drops as if dead, but with head alert, eyes challenging, ears cocked for a word of command.

"Guay to me, guay-da-me, Nell." Back around the flock, this time forcing the assemblage together down the hillside toward the gate.

"Doon, Nell."

The shepherd walks up to the flock. A series of short commands again move the dog around the flock, balling it into a tight mass roughly fifty feet in diameter.

"Coom by, Nell." The dog moves forward in a slow stalk, a lynx-like control of movement. A lead sheep near the shepherd moves out from the mass, passes by the shepherd. The shepherd counts. "One." The dog eases nearer. Another. "Two." Then the next. Like a thin stream pouring from a bottle, a single file of sheep siphons off from the flock.

"227, 228, 229, 230. Guay to me, Nell." The stream halts. Two groups of sheep exist in the field, the shepherd in between.

A command. The right-hand side of the larger flock is pressured, directing the group toward the gate.

"Guay-da-me."

"Coom on. Steady noo."

"Doon, Nell."

"By-da-me, by-da-me, Nell."

"Guay to me, Nell. Nell, guay-da-me."

Ninety sheep are still in the field as the last of the major group funnel through the passage. The shepherd walks through the gate, closes it. The sheep ahead are moving out. "Guay-da-me, Nell." The dog cuts them short, and they stop. Motionless, in a tight cluster, stemless carnations on a field of green.

The entire operation has taken seven minutes. The shepherd's name is Bill Elliot. A good shepherd, a good dog, out doing the same work as other shepherds are doing on hills all across Scotland.

The sheep-dog trials is an annual event held in many townships throughout Scotland. The shepherds come with their dogs to compete against other dogs and shepherds in the management of sheep through an obstacle course.

A group of three or five sheep are released three hundred yards distant from the shepherd standing with his dog at a pen. The shepherd must then maneuver them, by commands to his dog, through a series of gates down, finally, into the pen. He then releases them from the pen and the last one out must be split off from the others and sent in a different direction. Errors enroute, or taking more than ten minutes for the course, disqualify the pair.

It is a contest, but like the best of these, is based on a skill of one's daily work. Because of the distances whistle calls are used instead of voice, as in the hills, but the movements of the dogs are the same as by voice command.

The oldest trials in Scotland are held in the village of Yetholm, and for two days each fall the community is filled with shepherds and their talk. The White Swan Inn is a center of activity. In the evening the little bar is crowded with shepherds. It is a festive atmosphere, and two bartenders keep busy pulling pints of bitter ale and lager. Somehow amid the jostle of bodies enough space is kept clear for a dart game that goes on and on. "What's your score, shepherd?" someone calls. There is much joking and banter.

Some produce is exhibited at the Yetholm Show, but it is a shepherd's fair. The event takes place on the Haugh, a wet meadow and ancient place of common grazing. It is a day of wild weather. Gale-force winds whip across the ground and it snows briefly. The sun comes and disappears again behind racing clouds. Inside a tent, which has to be stitched twice to repair wind damage, two quiet judges appraise a display of lambing sticks.

Outside, a boys' band of pipers plays to the afternoon crowds who can buy tea, meat pies, and beer. On the far side of the judging pens where prize sheep were exhibited in the morning, clusters of shepherds stand. They wear gabardine coats and tweed caps and carry polished lambing sticks. Awaiting their turn they watch a far-off dog move sheep through a gate as a shepherd standing alone by the pen works patiently with him. Sixty-five shepherds have entered the competition and a new band of three sheep is used for each contestant. Early in the morning Bill Elliot has driven two hundred sheep from his farm to supply this need.

Now it is his turn, and he and Nell walk out to the pen as a white flag is waved signaling a keeper at the far end of the field to release the next group. The dog does well, but the sheep balk at entering the pen. It is very quiet, the movements of sheep and dog are slow-motion as man and dog *will* the sheep into the enclosure. The gate closes. Nell has won before, but today the prize is not hers. Other dogs have done better.

It ends at dusk and the Shepherd's Cup is awarded for best time and score. Someone pours a bottle of Scotch into the cup and it is passed from shepherd to shepherd. The judges, who have observed sixty-five dogs work during the day, drink. It is over until another year. It is just twilight as the two hundred sheep troop home again along the road, Nell holding them in control. The day has been filled with excitement and moments of tension, but it is nothing more than what is done in other fields by these same pairs 364 days of the year. The dogs are border collies, bred for this work and named for this region. They are black and white, and small, but most visible in their enduring speed and energy. One can almost smell the intensity of their surveillance of the sheep. Even penned, while they await dipping, the sheep see two eyes looking in on them, the dog lying alongside the closed gate peering through the slats, frequently pacing its length. The latched gate cannot open, even the dog knows that, but guards and looks in anyway. It is almost a taunt.

The quiet dignity of the sheep-dog trials, and the seriousness of the relationship between man and dog which is fundamental, is nowhere seen more clearly than in the ultimate words of reproach from man to dog. Frenzied by the stubbornness of a misdirected sheep, occasionally a dog charges in, nips, and by overreacting terrifies the flock which then bolts. A brief chaos.

"Doon, Roy!" The words like cannonade.

And then, more quietly, "That'll do, Roy." At the terrible words the dog flattens further as if to become invisible behind a shield of surrounding grass.

Pastore: the word is as soft as the landscape. Sheep-tending in a web of shadows cast by pale-leaved olive trees. The colors are soft, as soft as the sounds of the grove and of the liquid language. The movements are soft as the bicycle makes a pleasant journey to the sheep, and father and son share the daily vigil.

Even memories of the war are softened: "The front, it was right here. The Germans were disciplined and well organized; the Americans were always having parties."

Old farmland, once terraced for crops, is now lush in grasses. For centuries it has been owned by the Bourbons. Now it is rented for grazing. The hard lines of the holding walls smoothed by erosion, hidden by moss.

Only the olive orchards are still pruned and harvested. One can sit under a tree while the sheep graze. One can doze, one can dream. The sod is always damp, so dry reeds are gathered for a seat. The shepherd can lean back in the warmth and coolness. Good dogs create invisible fence-lines that contain the flock. For an hour or two when the sun is highest even the sheep lie down and rest.

Man has made and eaten cheese for over six thousand years. We do not know the details of its discovery, but it is likely that it happened in this way:

Rennet—the substance most commonly used to turn liquid milk into a solidified clabber—is an enzyme found in the stomach of young nursing mammals and in some plants. The dressed linings of these stomachs have been used traditionally as containers for carrying and storing liquids. At some point an uncured lining was used to hold some milk, and was either set aside for several days or placed in the sun. Time and/or heat and the rennet turned the milk into a soft, raw cheese, to what must have been the amazement of the shepherd. In parts of the Mideast cheese is still made in this way—formed in goatskins.

Until fairly recently, the treatment of the stomach membrane and rendering of rennet was part of the shepherd's work. In most places today, however, rennet is extracted and made available commercially.

Cheese is nothing more than milk solids, but the varieties are endless. The cheeses we know best come from the milk of the cow, goat, and sheep, but they are made also from that of the yak, camel, reindeer, and horse. Cheeses, in fact, have been made from all the animals that man pastures. The milk can be fresh or sour, or a mixture of the two. The milk may be whole, skimmed, or enriched with cream. Buttermilk can also be added. Many cheeses are made from combining the milk of different animals, as well.

The milk may be raw, pasteurized, or boiled. Each will produce a cheese of different consistency and taste. Cheese can be either hard or soft, or in between. It can be eaten fresh or it can be aged. Cheeses meant to be aged can also be eaten fresh, and vice versa, and these cheeses have names of their own. Once made, the cheese may be spiced, marinated, smoked, or flavored with herbs.

All of these variables are in turn affected by how the clabber is formed: by rennet, by heat, by bacterial action, or by a combination of these. The degree of heat and the amount of time the rennet is left in the mix before the curd is cut will change the cheese. The rennet can be made from different animals or from certain plants. The size of the pieces into which the curd is cut, and the amount of pressure applied to remove the whey, will affect the final cheese.

From all the above alternatives come the cheshires, bries, fetas, roqueforts, caciocavallos that we buy today. Many are named for the place where they are made; others are named for what they are made from.

Twice a day, every day
for eight months after lambing, Giuseppe Nesti milks his sheep.
Following each morning and evening milking cheese is made with the
assistance of his wife, son, and daughter-in-law, as well as grandchildren.

The process of cheese-making is no longer accidental, but neither
is it much more sophisticated than it was centuries earlier in these
Tuscan hills. Pliny wrote praise for the pecorino of the region. Together
with ricotta, it is the cheese that is made in this household.

The milk is first poured through a cloth to remove any foreign
matter: dirt or strands of wool from the sheep. Then it is heated, and
rennet added. Today, some shepherds use a thermometer to determine
when the desired temperature has been reached, but others still test by
spilling some drops on the upturned wrist, like a mother with her
baby's bottle.

Suddenly the marvel occurs. The liquid forms into a solid. This is then cut, stirred, and cut again. Soon a division is visible between the curds and the whey. When the separation is more or less complete, the curds are scooped out of the vat and pressed into molds. A board constructed as a shallow chute is used for working the cheese. By pressing and repressing the cheese into the mold, the whey is forced out and runs down the chute into a large pail. The cheeses are removed from the molds, turned over, reinserted, and pressed some more. The process continues for several minutes until the curd is quite firm and dry.

The whey that has been extracted is then reheated, along with additional fresh milk that has been kept aside for this purpose. Presently, new curds are formed. These are removed and gently pressed into forms. This is now ricotta, ready to be served.

The cheeses are then taken to a cool, dark room to ripen. The ricotta will be sold to a few local shops the next day, but the pecorino will be left on boards to age. People will come to this shepherd to buy some of the cheeses. The remainder will be sold at the market. The more aged the cheese, the higher the price. Pecorino to be used for grating will be rubbed with a mixture of dregs from the olive-oil keg and flour, and left to age until it is hard. An average of eight kilos of pecorino and ten of ricotta are produced daily inside the small farmhouse.

While the milk is heating, the evening meal is taken: great slices of bread, spaghetti with thick green olive oil from this land, grated sheep cheese. And after the cheese is made, some ricotta, still warm, is eaten with a little sugar.

The cheese is made, the day is over. The old man is tired, stirs and stirs his cup of coffee, smiling, half-aware of his surroundings.

Sardinia is also a cheese-making region of Italy, but with a different temperament from that of Tuscany. It was here that *la razza sarda,* an excellent breed of milking sheep, originated.

Today, shepherds watch over their sheep with great vigilance, for sheep-stealing is endemic to Sardinia. In a recent nine-year period over a hundred thousand sheep were stolen. At night the shepherds sit wearing black hats and black capes that they wrap completely around themselves. If they smoke, they do so by placing the burning end of the cigarette in their mouths so that no light is visible. They wait with their guns for the sound of sheep bells that signal disturbance. Often, however, it remains quiet as skillful thieves stuff grass into the bells, silencing the clapper.

The traditional shepherd's dwelling in the hills is constructed in tipi fashion. It is of massive dimensions, the door itself measuring three feet by six feet and weighing a hundred and fifty pounds. Inside, some sheepskins that are used as bedding are rolled and hung up during the day. A chunk of cork forms a simple stool. The fire is made on the dirt floor, near the center of the space, and the entire interior is burnished with a patina of smoke.

The same sheep, almost. The same grass, almost; pastures separated by only a stream or hillock. Each shepherd making cheese in the same way, almost; cheeses ripening in dark cellars of the same cool temperatures, almost. But each cheese is stamped with a particular mark or set of initials, making it "different" from the others.

If there were no mark we might not know from which shepherd and flock a cheese came. But the shepherd could say: it is mine. As has been written about the farmer's furrows on a field: although all were straight, all were different, and a man could read them. "It was his signature, not only on the field but on life." And to eat a cheese tomorrow will be to remember that.

Not more than sixty miles from the ruins that were once the splendor of Knossos there are mountain pastures. Here, on this small island that spawned such a complex civilization and rich, noetic art two thousand years before Christ, the last of simple herding traditions are holding on the longest.

The stani *(sheepfold) belonging to Miltiades Xsilouris is located twenty-five kilometers from Anoyia, under the eight thousand-foot majesty of Mount Ida. Until a year ago this place was accessible only on foot, with donkeys the beast of burden. There is no power or other municipal infringement.*

In fact, nothing has been added. Some stones have been reorganized with skill to form a series of round enclosures, some for beast, some for man. Only two enclosures are roofed, again with stone. Within the smaller, cheese is made. A second, slightly larger, is used for sleeping and storage. Blankets thrown over a thin layer of cut branches cover half the dirt floor; elsewhere there are a few tools, a five-gallon jug of wine, some food. A third enclosure has a loose web of sticks crisscrossing overhead to provide shade. Here, on stone benches, the shepherds sit to eat, smoke, relax. Overhead, boards of cheese ripen in the soft wind. Adjacent to this complex stretch long elliptical corrals.

Except for these formations of rock, when the shepherd and his flock depart there is no sign that man has ever been in this place. Yet five shepherds live here, share the brush bedding, and the food.

The extreme generosity of those who have little is exemplified by this shepherd—proud owner of many sheep, head of family, hard worker and, in this place, the host.

If a stranger passes by Miltiades's place a baby lamb or kid is soon roasting on the fire. Cheeses ripening for the market are taken from the drying boards to be shared. Wine is poured. One cannot leave before accepting this hospitality. Friend, relative, stranger—all are guests.

MILTIADES XSILOURIS «VOSKOS» ANOYIA

Anoyia is an hour's drive from Knossos along a winding mountain road.
It is a small town terraced into the harsh slopes of a hillside on three
levels. A few wooden tables and chairs border the narrow, almost empty
streets. There is a quiet, almost lazy atmosphere as men sit, and sip at
pungent coffee, syrupy in its sweetness. Everything is whitewashed—the
whole town shimmers in waves of white. The interior of almost every
house is white, too, but cool and dark and with movement as women
dressed in black sit at looms weaving shepherd's sacs. They are woven
for husbands and also for the tourists who come here for the brightly
dyed cloths. A visitor might almost feel as if he were in a sleepy
Mexican village. But he would be wrong: Anoyia is a tough town.

Crete was dominated by the Roman and, later, Byzantine empires;
and, in 1204, was occupied by the Venetians, who subsequently ruled
the island for four centuries. Resistance to their presence was centered
in Anoyia and punitive expeditions were sent to the mountain town to
subdue opposition. In 1669 the Turks won control of Crete. Opposition
to the occupation again originated principally from this area. The
Turks retaliated against Anoyia.

Early in World War II the Germans overran Crete. Once more history repeated itself as resistance leaders established their headquarters in Anoyia. Then, on August 13, 1944, the German command issued a final statement concerning Anoyia: "We order the destruction of the village and the execution of any male inhabitant found in the village or in a distance of one kilometer around."

Today's residents of Anoyia, the survivors, among them the family of Miltiades, carry this history in them.

Milking a sheep does not take long, but with five hundred ewes the work takes about two hours. The entire herd is milked each morning and evening.

At first light the sheep are gathered and penned in the elliptical enclosure constructed of stone and reinforced at low spots by tangles of briar. There is one small exit in which a large round tub is placed. Behind this receptacle a board is erected to prevent sheep from walking into it. As a sheep comes by either side, she is pinioned between a shepherd's legs. Her rear is lifted and aimed toward the tub, and her udders are squeezed and pulled in a short series of powerful strokes. The milk arcs into the tub in hot spurts, the final drops are stripped from her teats, and in fifteen seconds she is dry. If Miltiades is the milker the sheep is dry in ten seconds. No one's hands and arms can compete with the strength and ability of his which look deceptively soft from continual encounter with lanolin.

After the milking the ewes are sent to graze in the choicest pasture. The rams are on poorer land and they are checked. The goats, which exist on land that is thick in brambles and weed and almost without grass, are also milked. The division of labor is established and regulated within the group by accepted custom.

Several meals are taken throughout the day, during periods when the animals do not require attendance. The food is simple: cheese and flesh of the flock. Male lambs that compete for the milk supply are killed early; barren ewes that consume an unnecessary share of pasture also will be eaten.

There is no refrigeration, and the animal must be cooked immediately after it is slaughtered. A portion of it will be eaten within the hour, the rest of it consumed within the next few days. Large scallions are eaten with the meat. There are no plates, just great chunks of bread. There is a clean acidic wine, *krassi*, made in the village. A moist mound of *misithra*, a soft cheese still warm from the making, is cut in sections. The saltier and more aged cheese made from the first pressing of curds is also eaten. After two months of drying this becomes *kefalotyri*, a hardened cheese that is more commonly used for grating.

Miltiades's uncle, Epaminondas, finishes eating the meat from a shoulder blade. He carefully scrapes it clean. When held to the light, the center becomes translucent and the thin bone structure creates a design within the light. From these abstract shapes the shepherd reads fortunes, like a seer with a crystal ball. For the shepherds, the messages usually prove to be true. Most frequently they are a portent of weather, health, or adultery.

During the afternoon, while the milking is in progress, a shepherd arrives with his donkey which has been pasturing for some days on the hill. The animal is thin and has not been eating. The herder calls out to Miltiades that the donkey has an abscess on the roof of its mouth that must be treated.

A pick-axe is produced, and after the donkey has been tied to a tree its mouth is forced open and the handle wedged far back, locking the jaws apart. The shepherd takes his knife and makes a deep cut through the infection into the roof of the mouth. Then, placing one hand on the donkey's nose for leverage, he scrapes the roof of the animal's mouth with his fingers, stripping the infection from the area. He continues this until the blood that is draining is clean. With the same knife he cuts an onion and mashes it into the wound. Finally, he takes a great handful of salt and massages it vigorously into the cut. The bleeding stops, the pick-axe is removed, the rope untied. The donkey grinds its jaw a few times, gives two "haws" and walks away slowly.

The shepherd washes his knife and hands, and comes over to watch the last score of sheep being milked.

Approximately 275 pounds of milk are produced daily between the months of January and July. During the latter part of June the rams are put with the flock and the sheep dry off. The season of hard work is then over.

They look like warriors. Tall, proud men with black mustaches and dark eyes. The visage is accentuated by the line of a black bandana, the traditional headpiece of the Cretan shepherds. High black leather boots, broad belts, and long knives.

Warriors without glamour or contest. Shepherds. A company of men who spend their day milking sheep; men who in former days would have seen their wives only rarely—moments of procreation not of romance. Men who, with modern access to the village, choose still to remain among themselves. A company of men constructed on traditional familial lines. A hermetic cluster in the hills.

It is not a quiet or solemn group. During periods of work and relaxation there is a continuous banter. These are men that express momentary feelings, but for whom, perhaps, other emotions are muted within a hard life.

The transhumance for the shepherds of Crete is more complicated than it is for their French counterparts. While the latter simply make the trip and keep the sheep on the trail, here the shepherd must also milk his sheep twice a day en route and then make cheese. This is true in all of Greece. Writing about the Sarakatsani herders on their journey to the Zarobi Mountains, J. K. Campbell described their transhumance in this way:

This is a moment almost of festival. The sheep again carry their bells, which are not worn in the winter. As well as their practical function of identifying the flock they express the joy of the *stani* that the difficult winter period is over and its pride in the appearance of its flock. But the journey itself is not pleasurable for the shepherd. Particularly in recent years it has become difficult to find grazing on the road and there are hostile and sometimes violent encounters with villagers and agricultural guards. The sheep travel eight to ten miles each night on a road which is often congested with the flocks of other *stanis*. In the dark, and on ground he does not know, the shepherd must try to graze his sheep where he can, prevent them from mixing with the animals of other flocks, and guard them from village . . . thieves for whom the conditions are an opportunity and a temptation. At dawn they try to find some grass, and here the mule-train with the older men, women, and children overtakes them to prepare a temporary milking-pen. After the animals have been milked the women at once begin to turn the milk into cheese, and later to bake bread. Meanwhile the shepherds graze the sheep until about 11:00 A.M. when the animals lie down to rest in the heat of the day. They are milked again in the evening before setting out on the next leg of their journey. But the women remain to make the cheese for the second time in the day, and to take a few hours of sleep before they move swiftly through the night to overtake the sheep once again.

What in summer is a dry sunny landscape will be lost in impassable snow in November. The autumn transhumance is a descent back to the flat, near coastal section of the island. It is a region of cultivation— of olives, raisins, almonds, and honey bees. Wild daisies and poppies color the terrain. Miltiades leases four thousand acres of pasture here. He and the other shepherds of his *stani* live in a small building that was recently constructed. It is a long way from Anoyia, but only a fifteen-minute walk to the small village of Pombia. A miniature taverna becomes a meeting place for the many shepherds from the hills above Anoyia who winter their sheep in this area. It provides an alternative to their daily diet of lamb and goat. Here they are served skewers of pork. It is also a place of wine and, sometimes, music.

It is the time of year for shearing. There are several hundred sheep to be clipped. The hills are dotted with the *stani* of other shepherds, many of them relatives, and the shearing becomes communal work. The previous week an uncle's flock was done; next week it will be a cousin's.

The day before the shearing is to begin, eight lambs are taken from their mothers and dropped into a small stone hive. A pile of bramble is stuffed into the roof opening to prevent escape. The enclosure resounds with bleating, and eight mothers circle it and give voice in return. The morning milking is finished, cheesé is made. The sheep are pushed out to graze. There is a general neatening of the complex of huts and pens. Meals are eaten, and the day again becomes routine, except for the incessant bleating of sixteen sheep.

At midafternoon while the shepherds are gathering the flock for the evening milking, women of the family arrive from Anoyia. They are not often seen here but are welcomed now. The sheep are penned and the men, as usual, do the milking. An occasional sheep is grabbed, its teeth examined for age, and after being milked is led to the back of the *stani* and slaughtered. Afterwards, the flock again returns to pasture, and the women sweep the entire area with handfuls of coarse brush. Then as night falls they return to Anoyia.

The killing and butchering of sheep or goats are done with great speed and precision. One cut with the knife and the animal is unconscious from loss of blood in ten seconds, dead in thirty.

A short slit is then made on a hindleg and a stick inserted in the hole and pushed along the leg into the body. This is moved rapidly up and down several times to free a passage between skin and flesh. The stick is then removed and the shepherd starts blowing into the hole. The casing of skin and fleece begins to separate from the carcass, growing larger and larger with each breath until it resembles a huge freeform balloon. The shepherd then whacks and squeezes the affair to loosen the skin from any remaining contact with what it protects. With a very few number of cuts the skin is free and is pulled off.

A single long cut from tail to throat is made and the entrails are removed. Almost everything is kept for eating. The small intestines are pulled free, stripped, and coiled like a rope all in one series of motions.

The entire process takes under ten minutes.

The morning of shearing day. Work—gathering and milking—starts at dawn. More teeth are examined and a new group is added to the back wall. Family and friends of all ages are arriving with baskets of bread and other food. Some women are dressed in traditional black, others in party dresses. The shepherds are in their normal dress, other men looking more citified.

Once the milking is finished the work is divided among the many hands. Four men complete the butchering of the ewes and start on the penned lambs. Others gather huge piles of scrub for kindling. An uncle leaves by donkey in search of solid firewood.

The shearing starts. It is rapid work, casual without being careless. A mosaic of people and beasts: bent men, trussed sheep, women gathering armfuls of fleece to be stuffed into sacs, startled sheep racing to the hillside bleating in the surprise of new nakedness.

9:30. Women move through the confusion with pitchers of wine and bottles of raki—a fiery distillation of grapes—bread and cheese. A man rises from his shearing, still holding the shears in one hand, pours wine into the proffered cup, swills it around, pours it on the ground, then fills the cup, drinks it, cuts a slab of cheese, eats it, and returns to work. The scene is repeated. Periodically, a shepherd stops for a few moments to whet the blades of the shears. Although the pace of work continues, there is much joking and laughter. The group stops to watch a ewe dash off, its fleece dragging along the ground still attached by one tangle of wool. A shepherd grabs it as it passes and, with a snip, frees her.

10:30. The fire has burned sufficiently for cooking to begin. Foot-long sections of large intestines are blown up. Into each is threaded a spleen that has been cut in a coil to form a long section. Around this is placed a sheet of the filigreed stomach fat. The whole is then bound round and round with yards of small intestines. The finished item, *kokkorétsi,* is then threaded onto a skewer and set alongside the fire to broil. Epaminondas is the cook, and as carcass after carcass is brought to him he repeats this work, stopping frequently for a glass of wine, a bite of food. Skewers of lung, skewers of heart, skewers of liver.

11:45. Shoulders, legs and loins of lamb and mutton have now been butchered and are cooking. Most of this meat is broiled, but some sections are boiled in a huge cauldron.

The day has become hot; still the wine is passed and the first grilled skewers are offered. Hunks of the broiled meats are sliced and eaten with bread. Thick scallions are consumed.

12:15. The shearing is nearly completed. Only the black sheep are left, and soon they, too, are shorn, their wool placed in a special bag.

The work is finished, the food is cooked. It is time to relax. Beyond the nearest hill is a spring toward which the group moves, carrying the provisions.

Miltiades, who has been at work since dawn, is everywhere—laughing, talking. There is mixing of ages and of the women and some of the men but the shepherds remain largely to themselves, not isolated but apart. An endless procession of food makes its way through the assemblage. More wine is passed, *misithra* and bread are eaten, a cauldron of noodles mixed with sheep cheese is offered, more meat, and finally a crate of oranges.

2:00. Now there is music.
A lute and a *lira,* the Cretan form of violin, fill the area with
a lilting melody. A singer's voice joins the driving repetitive sound of
the music. Then a few people leave the rocks and begin the sober yet
happy, masculine yet delicate, almost formal movements of the Greek
dance. More people join until there is a circle of dancers. Slow
pirouettes, a polka-like ring, and the high kicking, boot-slapping move
for which the dance is known.

3:30. Almost suddenly it is over. Each shepherd must now make the
journey to his own hill, his own flock, for the evening milking and
cheese-making. The women descend again to the village. Everyone
is happy but is feeling a little tired and used by the long day of work
and eating. Miltiades stands by the spring until all have left.

It is time to go to work. The spring is deserted.

Lambing time is like the harvest. It is during this period that a shepherd will find out whether or not he will have a good year. Success or failure, profit or loss—insofar as these things are quantified—become fact at lambing. Drought and disease can be disastrous for a shepherd as they can be for a farmer with crops.

But lambing is also different from the harvest. Whether tending fields of wheat or grapes, turnips or potatoes, there are many points within the seasons when the farmer's energies are called forth to prepare the soil, plant, cultivate, and reap. For the shepherd, the comparable effort is concentrated in one time. Five months after the rams have been put with the flock, lambing starts. Then for a period of weeks the shepherd will hardly ever sleep. During lambing Yves Hébrard is with his flock from 1:00 A.M. to 5:00 A.M.; 6:00 A.M. to 11:00 A.M.; 1:00 P.M. to midnight.

In the Highlands and other unfenced spaces birth happens according to nature. But even there the shepherd walks daily into the hills, searching for problems among his flock lost in the expanse. He will not know, however, the extent of his loss or gain until he gathers them in at shearing time.

It is long, tiring work. Except on the moors, a few days before lambing the sheep are brought off the hills to specially constructed lambing pens or fields. In former times the protective barriers were simple, woven by the hurdle-maker and thatched with straw. They were set up and rearranged according to need. The shepherd moved his tiny hut to the field. In France this looked like an oversized coffin with gabled roof, in England a tiny gypsy caravan measuring a mere eight feet by four feet by five feet, which was mounted on wheels of miniature dimension. In between brief snatches of sleep in his hut, the shepherd would go forth with a lantern, listening for a bleating that signified distress, looking for sheep downed with problems. Then grasping and tugging, he would maneuver a tiny wet thing to life.

Today it is still much the same. Although modern methods have replaced much of what was once done by ancient lore, the work of lambing continues basically unimproved.

The timing of lambing is not fickle: the lambs must be born just as the sharp edge of winter begins to dull, in order that they be weaned and ready to feed on the best spring pasture. Sleet and snow, storm and frozen ground usually greet their arrival. They may be born at any hour, but, like most creatures, it is usually at night.

The shepherd watching by his ewes and theaves
All night in loneliness, each cry knows well,
Whether the early lambing on the Downs
Rob him of Christmas, or on slopes of fell
March keep him crouching, shawled against the sleet;
But there's a cry that drowns
All else to shepherd's ears: the wavering bleat
Of weakling newly-born: then he shall lift
The lanky baby to his own warm hut,
Lay it on straw, and shift
Closer the lamp, and set the bottle's teat
With good warm milk between the lips half-shut,
Coaxing the doubtful life, while wind and rain
Against the window of the cabin beat,
And homing cottars in the plain below
Look up, and seeing the window's yellow glow,
Mutter, "The shepherd's at his job again."

Depending on the number of rams and the size of the flock the whole process will take anywhere from a few days to a few weeks. Nowadays a modest system to help organize the event has been introduced in some areas. As the gestation period is almost always an even 150 days, the shepherd can divide his flock into those bred on the fifth of the month, the sixth, etc. The identification of bred ewes, "tupped" ewes in the language of Scotland, is simple. Each day the shepherd smears a blob of paint or tar on the belly of the rams. At day's end he can pass among his sheep and, by the telltale sign of color on some rumps, identify and mark the group to lamb 150 days thence.

Although this gives order, it does not reduce the work. Several ewes will lamb within the same short period, and, as with all births, there can be complications. There will be lambs born dead, ewes will die giving birth. There will be lambs and ewes that would die if the shepherd were not there to give life. There are some lambs that will die anyway, in spite of all the care that the shepherd can provide. If a lamb is positioned wrong in the womb it may rupture vital organs as it begins to move and so the shepherd, arm extended deep into the sheep, carefully untangles the lamb by feel, and guides it to life. But once born, no matter what the complication, in that mysterious moment of knowing as the mother turns to lick the little thing clean, a wobbly creature is already butting its head up under the udder to suck.

Sheep must be among the most casual of animals about giving birth. In fact they do not even appear pregnant. It is difficult for a shepherd to know if, in fact, they are until two days before delivery when they "bag up"—their udders start to fill with milk. Unlike many animals they make no preparations—nest, lair, bed—for the event, and they let their offspring pop out as they stand, then after a lick or two will resume their grazing. If they produce twins they may only bother to encourage one to its feet.

A ewe will not accept another's lamb.
This creates a problem, for in the course of lambing there will be a
number of orphaned lambs, as well as ewes that will deliver dead
offspring. Whether Basque, Scot, or French, shepherds everywhere
solve this problem in the same way. A lamb that has died is quickly
skinned, less feet and head, the process taking only a few seconds. The
fleece is then either tied or sewn around an orphaned lamb. The
shepherd then brings this deception to the mother of the skinned one,
who, after a hesitant sniffing, will usually be fooled by the ruse. After
a day the mantle may be removed, the adoption having been made.

It is at this time of year that the shepherd's totem—his crook—is finally employed. In some countries it is called by its function—"lambing stick." It is the tool by which wayward lambs are caught to be doctored if need be, disguised perhaps, and brought to their source of life. Where lambing takes place unaided, as on the Highlands, the shepherd does not need, and therefore does not have, a crook in his staff.

And so another year begins, the cycle is renewed, and the shepherd looks again at the weather and to his pasture and resumes work.

NOTES ON THE PHOTOGRAPHY

Most of the photographs in this book were taken under relatively complicated conditions: shepherds live in isolated areas; during the five-day transhumance in France I had to carry all camera equipment, sleeping bag, food, and clothing through rugged mountain passes and often in driving rain and sleet; Crete, Sardinia, and Tuscany were dusty and remote; the Highlands of Scotland were wet and remote. Clearly my equipment had to be sturdy and as compact as possible. On almost all trips I took two Nikon bodies and 35-mm, 50-mm, and 105-mm lenses, graded sky and yellow filters, two light meters, and two small Braun strobe units. For the actual hikes up to the shepherds, I took along half that amount of equipment and replaced it when I could or needed to on return to the villages. I decided to use only Tri-X film because of its versatility and performance in contrasty situations. I used D-76 as a developer except for the pushed rolls, which required Acufine followed by a few minutes in D-76.

As for the actual photographing, I have to say that aside from the hazards of weather and terrain, the shepherds themselves made my task easy and joyful. Hardworking, serious, simple men, pleased and I think a bit amazed that a desire to record their way of life would bring David Outerbridge and me respectively from New York and Paris, they cooperated fully. Their wonderful faces, not to mention the glorious pastoral scenery, begged to be photographed.

J.T.

page 10: *Genesis* 4:1–5.

pages 11–12: From *The Odyssey* by Homer, Book IX. Translated by Robert Fitzgerald. New York: Doubleday & Company, Inc., 1963. Used by permission of the publisher.

page 13: From "Albion's England" by William Warner, Book IV, Ch. 20. New York: Adler, 1971. (Repr. of 1612 ed.)

page 13: From Henry VI by William Shakespeare, Part 3, Act II, Scene 5.

page 14: From *A Shepherd's Life* by W. H. Hudson. London: Methuen, 1910.

page 16: From *Seven Pillars of Wisdom* by T. E. Lawrence, pages 199–200. Copyright 1926, 1935 by Doubleday & Company, Inc., New York. Used by permission of the publisher.

pages 16–18: From *In a Hundred Graves* by Robert Laxalt, pages 66–68. © Robert Laxalt 1972. Used by permission of the University of Nevada Press, Reno.

page 20: From *The Mediterranean* by Fernand Braudel, Vol. I, page 94. New York: Harper & Row, 1972. Used by permission of the publisher.

page 22: From *The Drinking Well* by Neil Gunn. London: Souvenir Press, 1978. Used by permission of the publisher.

page 23: From *Far from the Madding Crowd* by Thomas Hardy, page 174. London: Macmillan, 1974. Used by permission of the publisher.

page 26: From *The Land* by Vita Sackville-West, page 23. London: William Heinemann, 1926. Used by permission of Nigel Nicolson, Executor to V. Sackville-West.

page 67: From *A Tour through the Highlands of Scotland* by J. Knox (1787) as quoted in *Scottish Country Life,* page 124, by Alexander Fenton. Edinburgh: John Donald Publishers, 1976.

page 68: From *The Highland Clearances* by John Prebble, page 24. Middlesex: Penguin, 1969.

page 72: From *The Drove Roads of Scotland* by A. R. B. Haldane, page 44. Newton Abbot, Devon: David & Charles, 1973. Used by permission of the publisher.

page 118: From *Honour, Family, and Patronage* by J. K. Campbell, pages 28–29. New York: Oxford University Press, 1974. Used by permission of the publisher.

page 133: From *The Land* by Vita Sackville-West, pages 16–17.

page 142: From *A Shepherd's Life* by W. H. Hudson, page 351.

ACKNOWLEDGMENTS

The authors wish to thank the following: Anne-Marie Brisbarre and Ivan Kats for their assistance in locating the shepherds, geographically and historically; Artemios Cotsifakis, Manolis Karandinos, Michele Picciaredda, Magda Salvesen, Archibald McLellan, Isobel Herbert, the late Prince Charles of Luxembourg, and R. H. Armstrong for their help in the various locations; and, principally, the shepherds of this book for their unending generosity with time.

"A shepherd's life, properly understood, is the richest in the world."